AVIAN NECROPSY MANUAL FOR BIOLOGISTS IN REMOTE REFUGES

BY

THIERRY M. WORK, DVM

U. S. GEOLOGICAL SURVEY
NATIONAL WILDLIFE HEALTH CENTER
HAWAII FIELD STATION

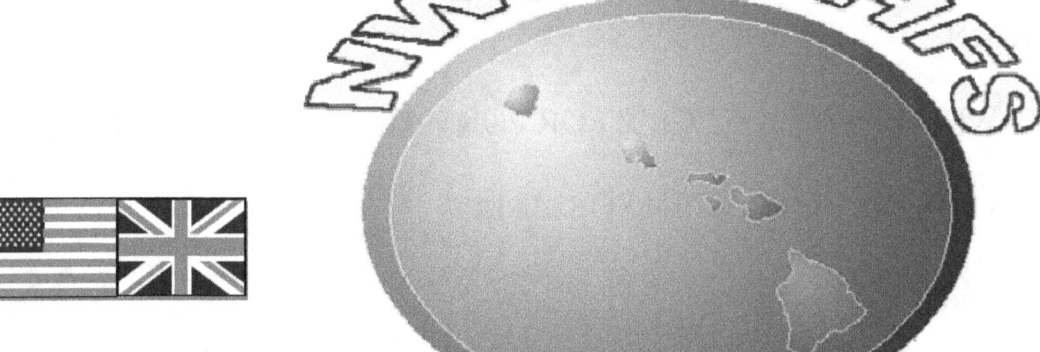

2000

TABLE OF CONTENTS

INTRODUCTION

This manual is for biologists in remote refuges who have little to no background in necropsy techniques. It is intended to assist you in recognition of bird organs and in procurement of appropriate samples for pathologic and other laboratory assays. The manual will probably be most useful in situations when wildlife disease specialists are unable to assist in sample collection due to remoteness or inaccessibility of the site.

WHY DO A NECROPSY?

A necropsy is one of the basic tools used to determine why an animal dies. It involves the thorough examination of a carcass externally and internally for any indications of causes of death (lesions). A good necropsy involves careful observations of lesions or abnormalities and procurement, labeling, and storage of tissue samples. Laboratory tests on properly preserved tissues allow wildlife disease specialists to systematically evaluate potential causes of wildlife mortality.

The better job you do with the field necropsy, the better the chance that wildlife disease specialists can determine what killed the animal. As such, select the freshest carcasses and, if at all possible, avoid freezing and thawing the carcass prior to necropsy as this can compromise microscopic appearance of tissues. When doing a necropsy, be observant and record your findings. If possible, take close up photos of interesting findings.

Generally, findings will deviate from normal either in shape, color, consistency, number or size. For example, a normal bird liver would be firm with sharp borders and have a homogenous chocolate brown color. An abnormality in the liver may manifest itself in the form of abnormal coloration (spots or blotches), consistency (too soft, too hard), size (excessively large or small), or shape (lumps, bumps or scars). Obviously, many of these interpretations require knowing what a "normal" organ looks like. Although this is best learned by doing many necropsies, reference to photographs (as in this manual), will aid the novice in assessing whether or not an organ appears normal or not.

MATERIALS NEEDED FOR A NECROPSY

Scissors	Toothed forceps	Rubber gloves
Plastic bags	Jars	Indelible marker
Knife	Cutting board	Water
Scalpel handle	Shears	10% formalin
Scalpel blade	Labels	Aluminum foil
Pencil	Paper	

Additional items that would be helpful include a scale, ruler and camera. Several types of plastic bags should be available including larger bags for carcass disposal and smaller bags (whirlpaks) to store individual organs.

The back of the manual has a recipe for making buffered formalin (a tissue preservative). It is unlikely that you will have the resources to make buffered formalin on site. An adequate substitute is mixing 15 parts of 37% formaldehyde with 85 parts seawater. **Placing organs directly in 37% formaldehyde or unbuffered formalin is unacceptable**.

SAFETY

When doing a necropsy, follow proper hygiene. At a minimum, wear gloves and do not eat or drink while dissecting a carcass. Remember, you don't know whether you're dealing with a disease transmissible to humans.

When working with formalin, **ALWAYS** use gloves, work in a well ventilated area and wash hands after all necropsies. All formalin containers should be clearly labeled.

LABELS

All labels should be written in indelible ink (e.g. sharpie) or pencil...**no ball point pens**. Minimum information on the label should include location of collection, date and unique specimen ID. To avoid confusion, abbreviate the month (i.e. MAR 5, 2000 not 3/5/00).

TAKING SAMPLES FOR LABORATORY ANALYSIS:

FORMALIN FIXATION (2 steps)

(Formalin fixation allows pathologists to examine tissues under the microscope and diagnose disease)

1) To ensure that enough formalin is present in the jar to allow for adequate fixation of the tissue, the ratio of formalin to tissue should be a minimum of 2 parts formalin to 1 part tissue by volume (Fig. 1). All tissues from one animal can go into one jar. **Label the jar.**

2) Ensure that tissue section is not too large to allow for adequate fixation. A piece of tissue should generally be no thicker than ~0.5 cm (1/4 in). If there is a lesion, make sure to take a portion of "normal tissue" adjacent to the lesion (Fig. 2). This is crucial as many diseases are diagnosed based on microscopic examination of the "margin" between a normal and abnormal tissue.

It is advisable to change the formalin once (say after 24 hours of fixation). This will result in better fixation and staining for microscopic analysis. Used formalin should be disposed of appropriately. **Tissues in formalin should never be frozen.**

FREEZING (1 step)

(Frozen organs can be used to isolate microorganisms or detect poisons)

1) Collect a good amount (20-30 g or 1/4 to 1/2 cup) of tissue, place in a small plastic bag, seal and label the bag using an indelible marker. In some cases, you may be asked to wrap the sample in aluminum foil prior to placement in a plastic bag. Collect tissue for freezing as early as possible during the necropsy to avoid contamination by gut contents, feathers, dirt, etc. Tissues should be placed in a freezer (-20C or colder is best) and kept frozen when shipped to the laboratory.

LID

**JAR WITH
FORMALIN AND
TISSUES**
**(1 part tissues
to 2 parts
formalin)**

FIGURE 1

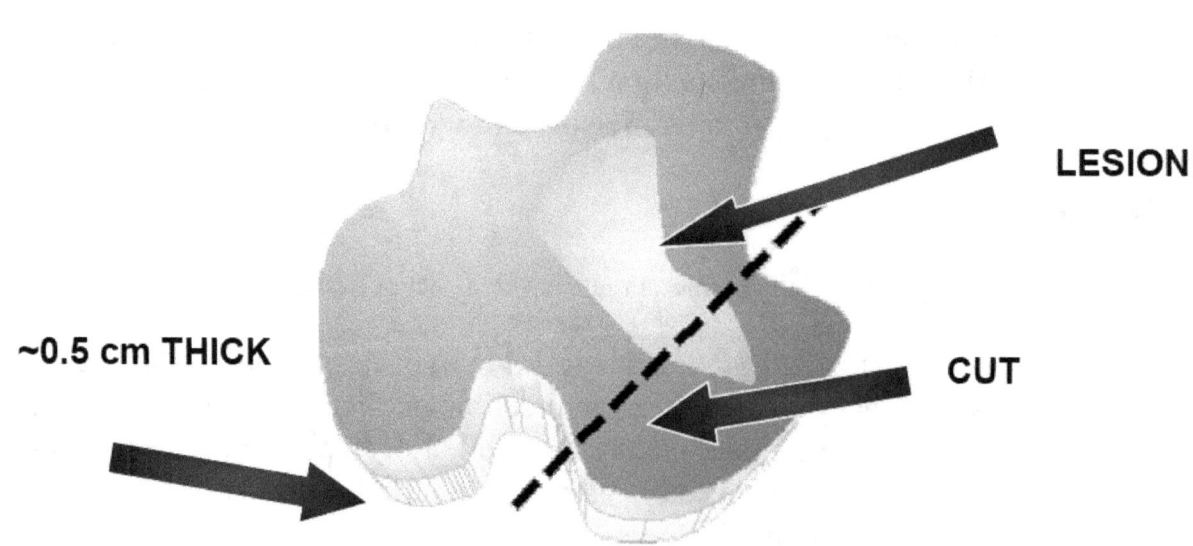

LESION

~0.5 cm THICK

CUT

FIGURE 2

THIS PAGE INTENTIONALLY LEFT BLANK

HOW THIS MANUAL IS ORGANIZED

The rest of this manual will show you, step by step, how to dissect a bird carcass using an albatross chick as our "model" bird. All birds have the organs shown here although size and shape may vary form one species to the next. The photos in this manual should give you a good general idea of what "normal" organs look like.

This manual is composed of a series of photos with a facing page of text. There are two sorts of icons throughout the text, scissors and glasses.

Sections with scissors icons are in bold type and describe the nuts and bolts of taking a carcass apart.

Sections with glasses describe organs and their appearance. Commonly encountered abnormalities appear in italics. Use these sections as a reference for taking notes on appearance of different organs. As you go through a necropsy, it would be wise to take samples of organs as you encounter them. There is a table at the end summarizing what organs you should have taken in formalin when you are done with your necropsy.

NOTE: This manual assumes you are doing a post-mortem on a fresh dead bird (either you saw it die or it died within the last 12-24 hours). Appearance of some organs (and their diagnostic value) will change dramatically depending on stage of decomposition so it is best to limit your efforts to the freshest specimens available.

Finally, remember TO NOTE AND RECORD EVERYTHING THAT YOU SEE. There can NEVER be too much detail.

EXTERNAL EXAM

Lay the bird on its back. Examine the bird externally from stem to stern for any abnormality or damage. You may want to take photos of any abnormality or for ID confirmation. When examining the carcass, check the following:

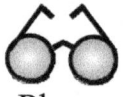

-Plumage: Is there down and where is it distributed? Do the wing, tail and body feathers look normal? Are the feathers clean?

-Weight of bird (if you have a scale)

-Major wing and leg bones and skull: are they intact?

-Cloaca: Is there pasting of feces around the cloaca and if so, what is the color? Is there anything protruding out of the cloaca?

-Nostrils: Is there anything (blood or mucus) leaking out of them?

-Mouth: the mucus membranes in the mouth should be pink. Colors like red or blue-gray are abnormal. Note any ulcers, cuts, plaques, growths, spots or lumps in the oral cavity. Also note the presence of foreign material or blood.

-Eyes: Are the eyes collapsed, cloudy, weepy? Are there abnormal warty growths or pustules around the eyes?

-Beak and feet: Are there abnormal warty growths or pustules on the feet or beak? If so, what is their distribution?

-Any other abnormality: lumps, bumps or exudates in unusual places.

Prior to starting the necropsy, wet the ventral (belly) feathers with soapy water to avoid masses of flying feathers and down during the necropsy. In this photograph, the bird has been soaked and a skin incision started on the neck.

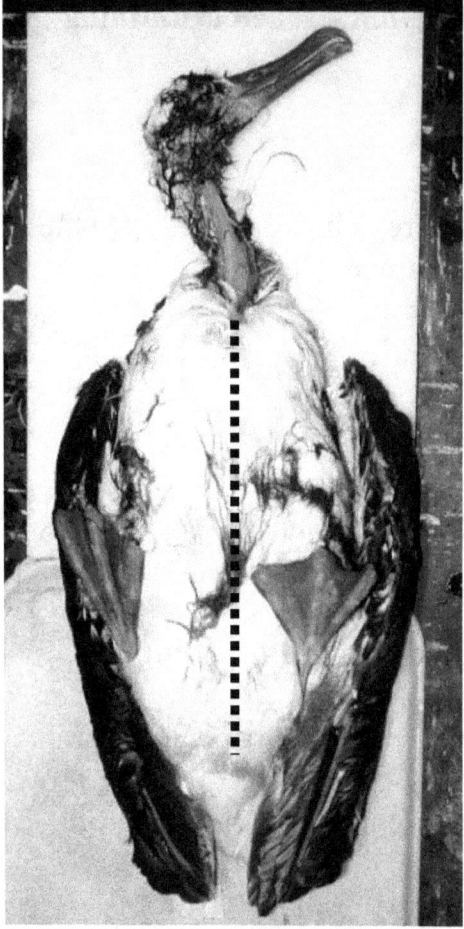

When you cut, keep the skin taut as it will make cutting easier. The best way to start is to cut the skin along the dotted line (as illustrated). Be careful with the scalpel blade when you reach the abdomen as there is only one layer of skin separating your blade from the viscera. Ideally, you want to avoid puncturing viscera early in the necropsy and contaminating the body cavity with gut contents.

Once you have cut down the midline, peel back the skin to expose the legs, breast muscles, keel bone and abdominal cavity as depicted in this photograph.

The following organs should now be visible:

TRACHEA (windpipe): This organ is a semi-rigid tube running down the neck parallel and close to the esophagus.

ESOPHAGUS: The soft tubular organ adjacent to the trachea.

BREAST AND LEG MUSCLES: The breast muscles are attached to the keel bone and should be a homogenous red brown as should be the leg muscles.
Abnormalities: in muscles include bruises, pale areas, or a gritty texture.

Now is also a good time to examine the skin for bruising which manifests itself as red splotches. If there is bruising, note its distribution. Also examine the neck to make sure it's intact.

Take your shears and cut the breast muscles and ribs along the dotted line illustrated. Note that when you reach the cranial end of the breast bone, you will encounter some thick bones that must also be cut (hence why we use shears). Once all the bone attachments are cut, you can cut away membranes attaching the keelbone to the body. These membranes are air sacs.

AIR SACS: They should be translucent to slightly opaque.
Abnormalities: Cloudiness or plaques on these membranes are abnormal and should be noted.

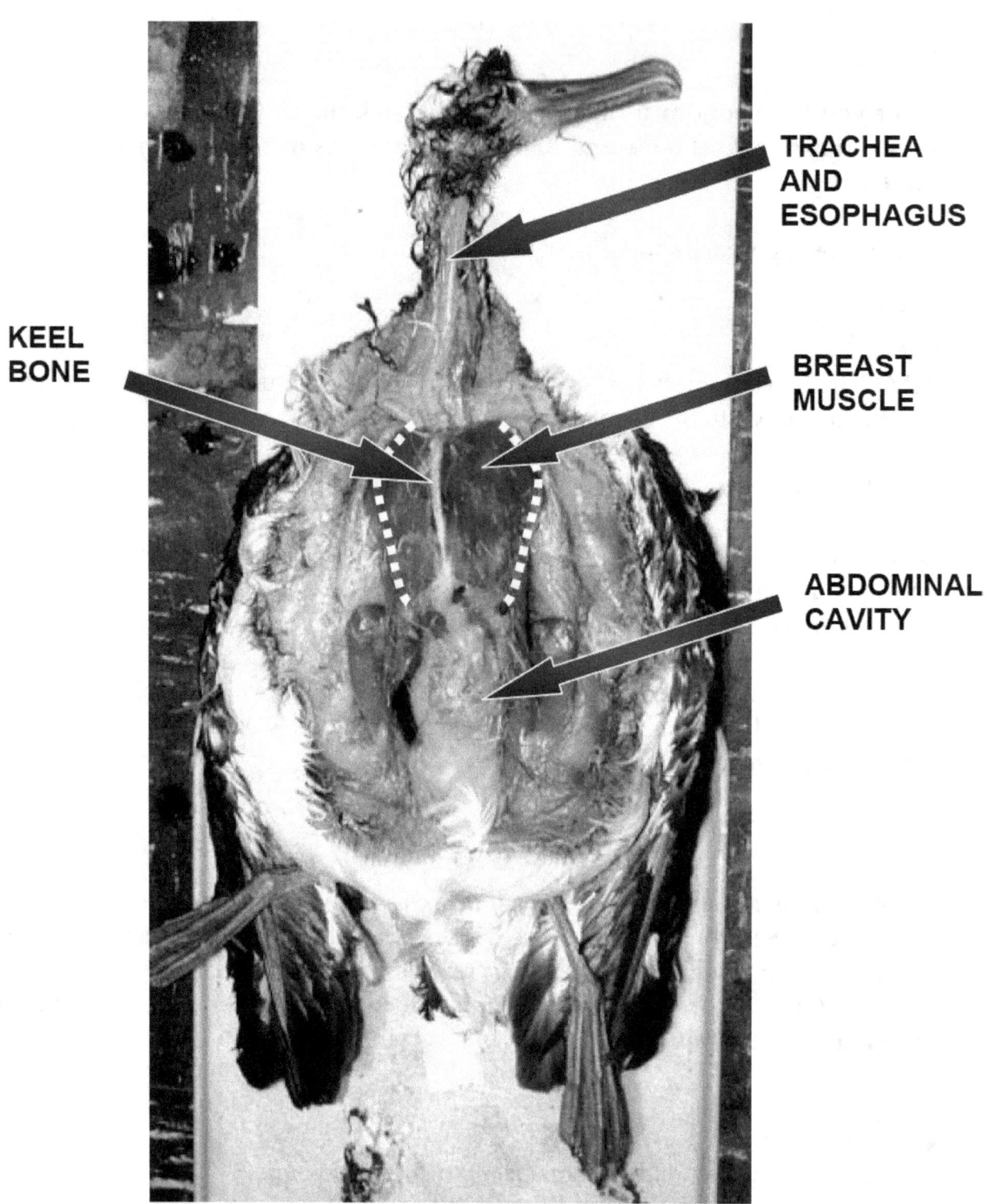

TRACHEA AND ESOPHAGUS

KEEL BONE

BREAST MUSCLE

ABDOMINAL CAVITY

With the keelbone and breast muscles removed, the following organs should now be visible:

ESOPHAGUS (mentioned earlier)

TRACHEA (mentioned earlier)

LIVER: Should be homogenous dark brown, firm, with a smooth surface with sharp borders.
Abnormalities: Swelling, rounded borders, rough surface, paleness, soft or mushy consistency, spots or a lumpy appearance.

HEART: should be red-pink and have a fairly smooth surface. The bird heart is similar to ours in that it has 4 chambers. Feel free to cut into it. The interior will have numerous glistening smooth ridges...this is normal. Also, in fat birds, there will be a line of fat on the heart near the head end. This fat should be firm and white or yellow-white.
Abnormalities: red or white spots on the heart muscle, a rough sandpaper like surface on the exterior or interior, semiliquid fat on the heart.

THYROIDS: These are small, round, pink structures near the thoracic inlet (where the neck meets the thorax). Note if one or both appear abnormally enlarged.

PROVENTRICULUS: The proventriculus is an extension of the esophagus and is a food storage organ. Because of this, it has in seabirds tremendous capabilities of distension. In this bird, the proventriculus is filled with squid beaks and plastics and looks quite distended. In birds with an empty proventriculus, it may be no wider than the neck esophagus.

Close ups of the above organs are shown on page 15.

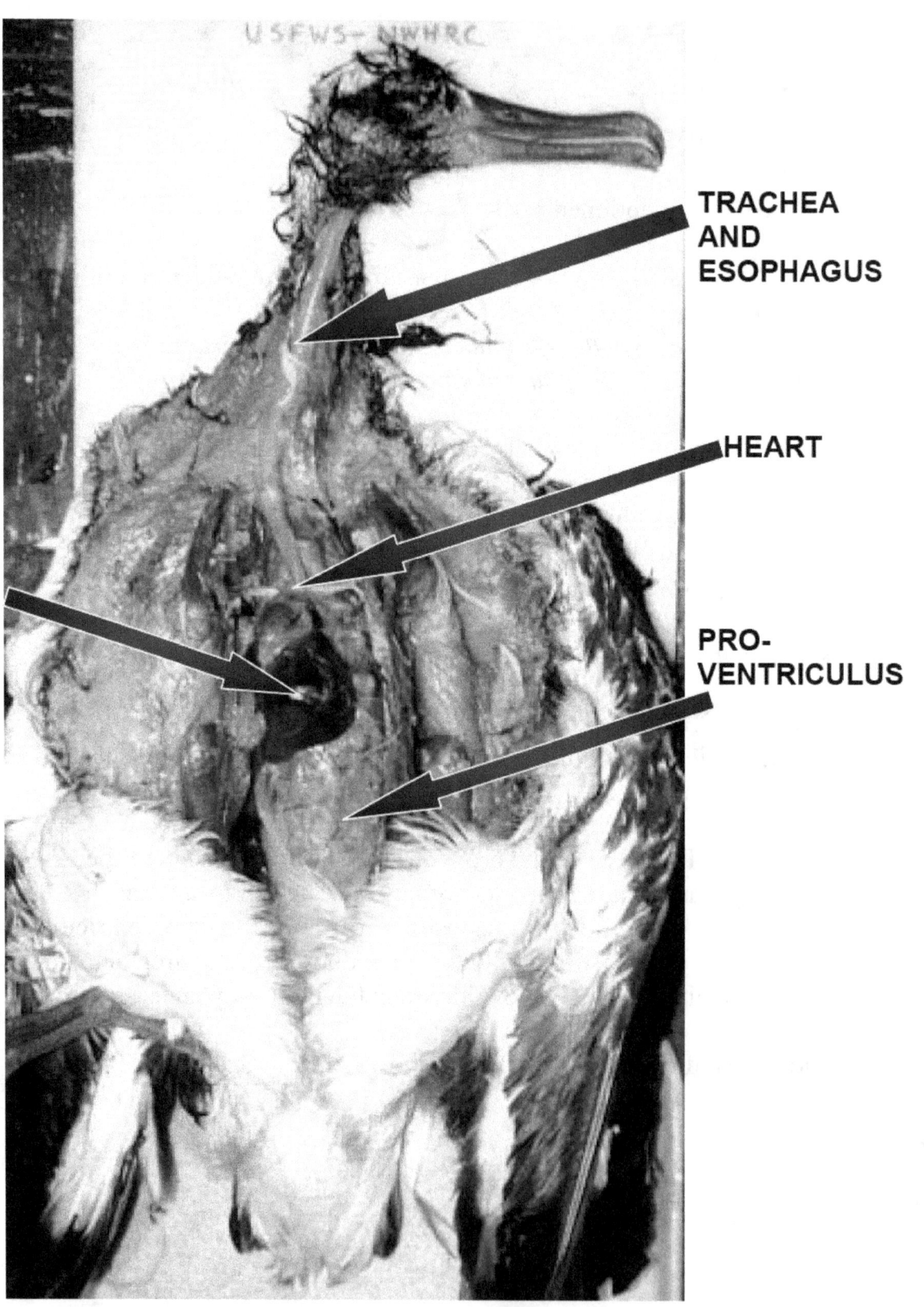

TRACHEA AND ESOPHAGUS

HEART

LIVER

PRO-VENTRICULUS

HEAD

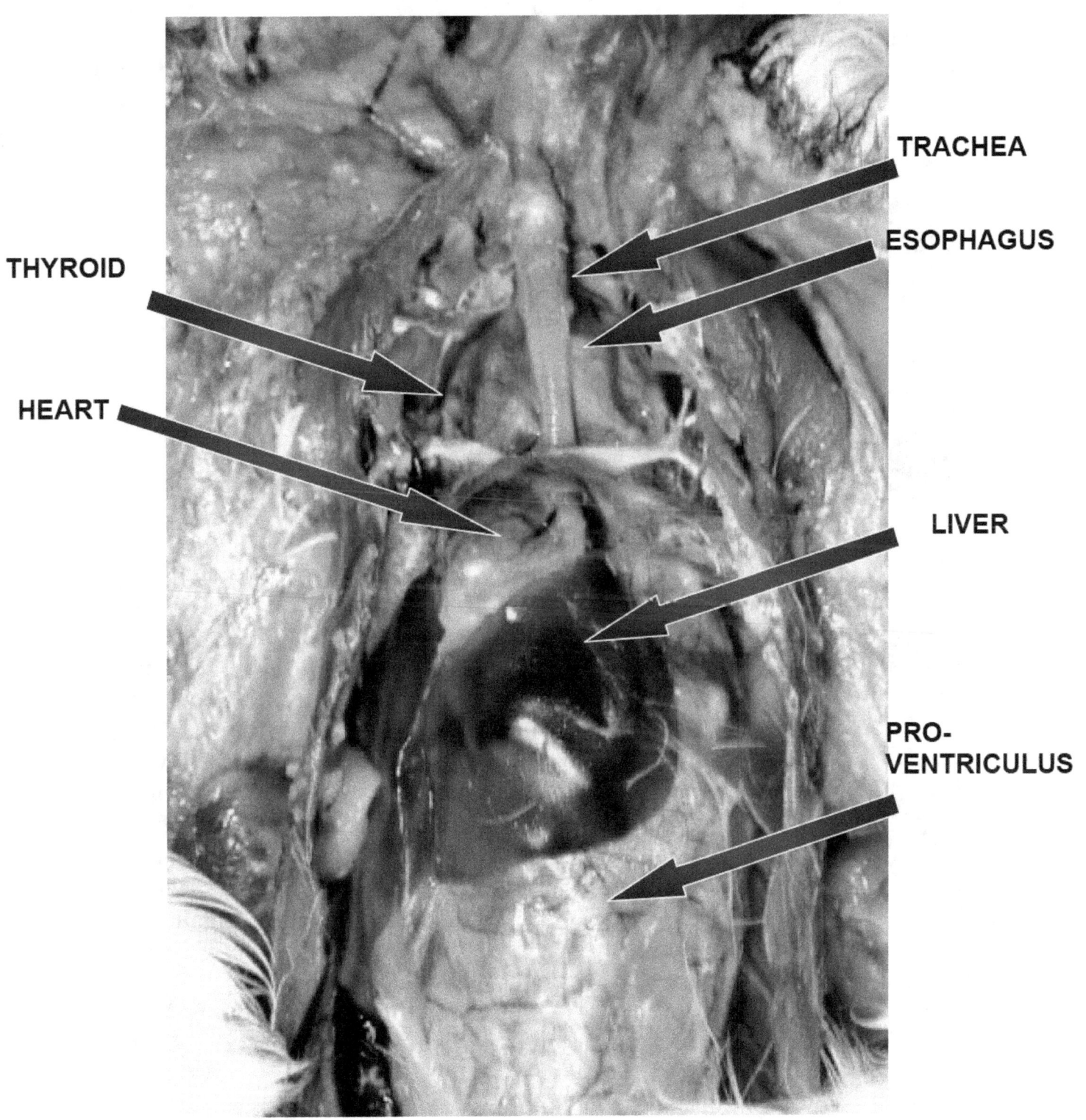

TRACHEA

ESOPHAGUS

THYROID

HEART

LIVER

PRO-VENTRICULUS

When removing the heart, grab the large arteries with your forceps and cut the arteries with scissors. Try to avoid grabbing the heart muscle with forceps as you can damage the tissue.

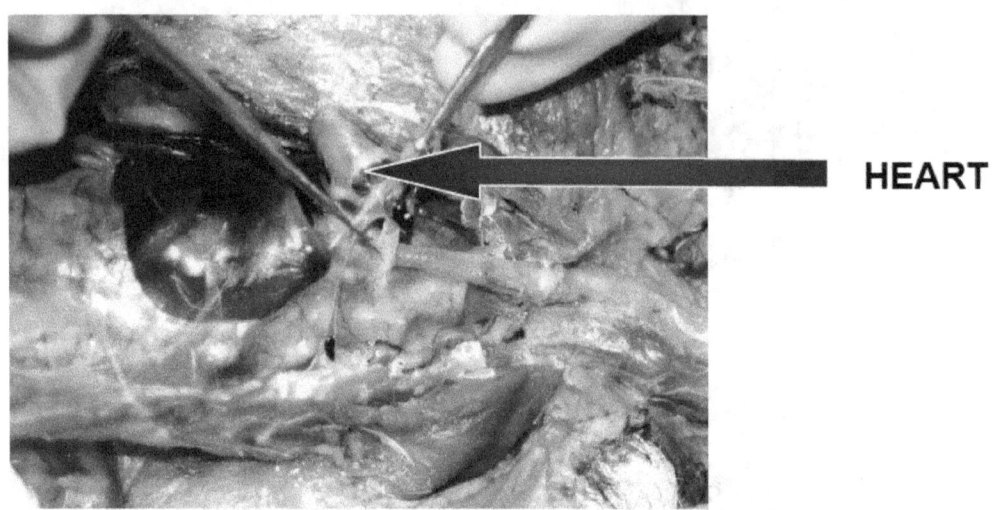

HEART

When removing the liver, try to hold onto it using the ligaments that attach it to the viscera. This is easier than grabbing the liver itself with forceps which will damage it. The other option is to gently hold the liver with your fingers. Again, try to handle the organ as little as possible. In some instances, when dissecting out the liver, you will see a sac distended with dark green material. This is the gallbladder which contains bile. Try to gently dissect this organ away without cutting it. If you do cut the gallbladder, try to limit the amount of bile contacting other organs as this will degrade the microscopic appearance of the organ.

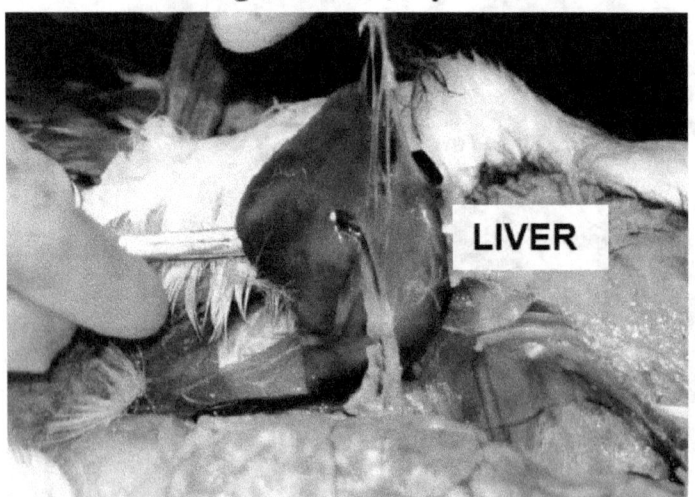

LIVER

This photo shows the carcass without liver and heart. In this bird, the proventriculus is massively distended with material (plastic).

HEAD

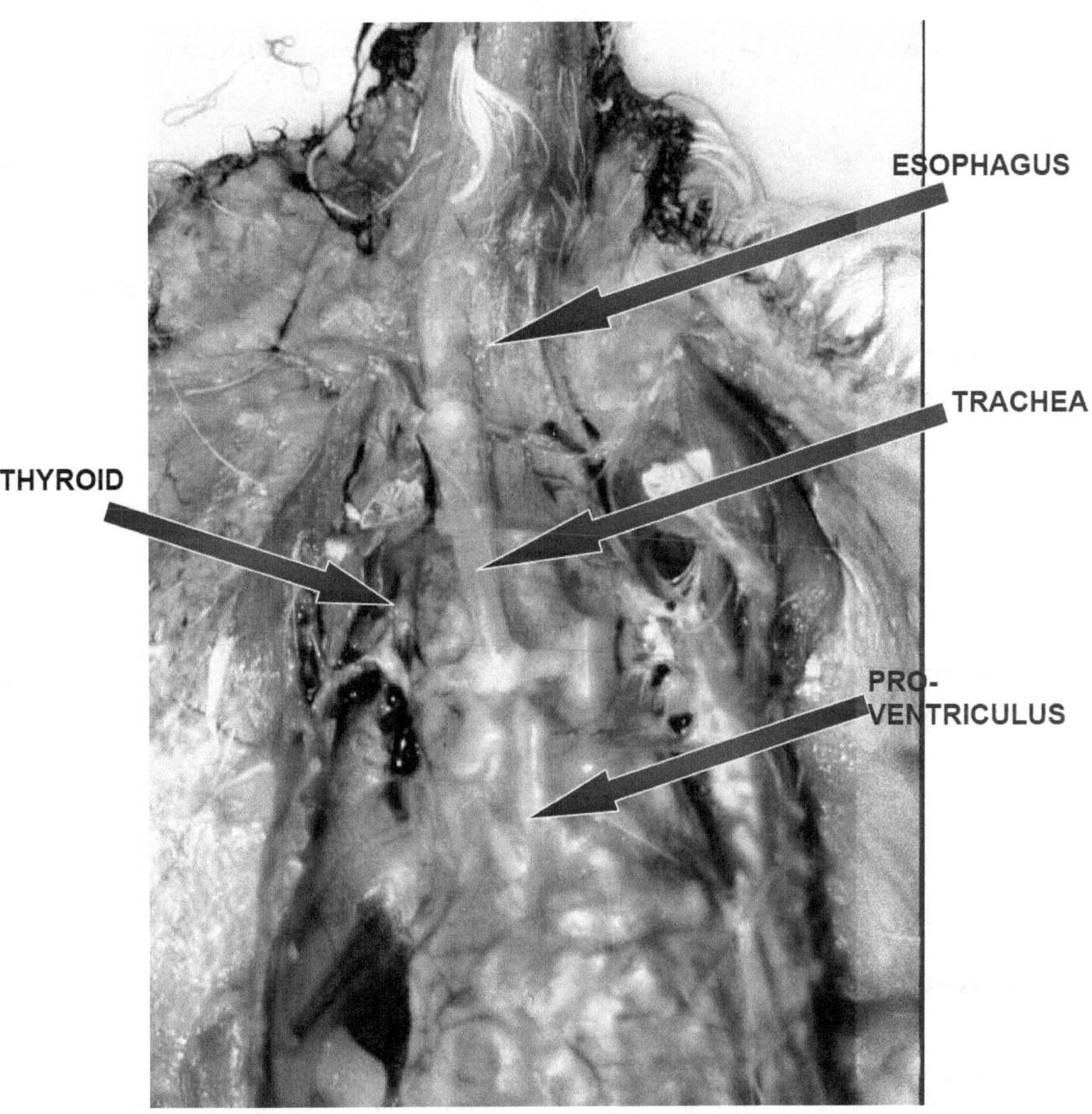

ESOPHAGUS

TRACHEA

THYROID

PRO-
VENTRICULUS

By rotating the proventriculus on its long axis, you will reveal the spleen located at the tail end of the proventriculus.

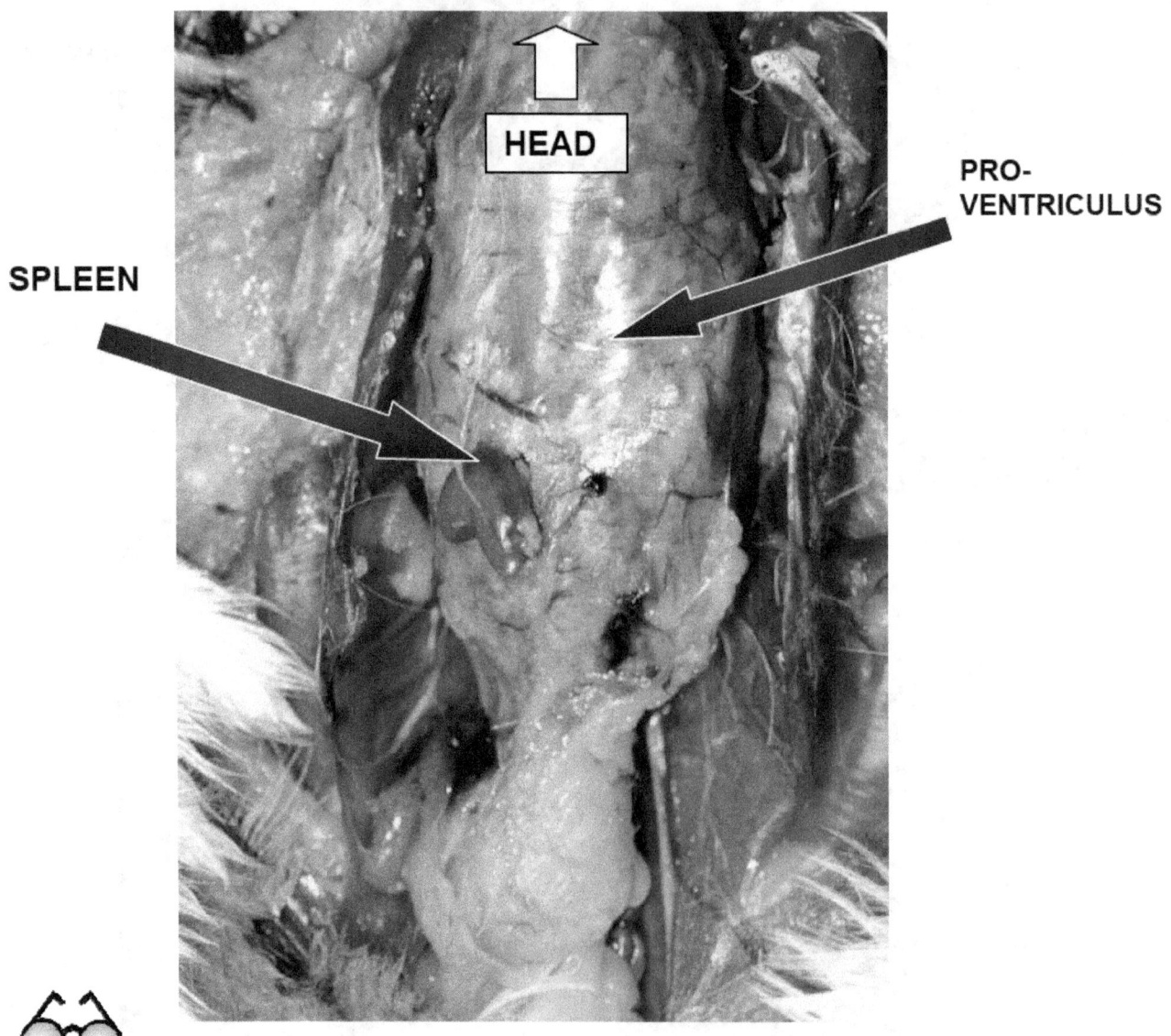

SPLEEN: This is a bean shaped organ located on the underside of the pro-ventriculus. It is normally brown-pink and about the size pictured here. *Abnormalities: A grossly enlarged spleen, nodules, or abnormal spots or splotches is abnormal and should be noted.*

REMOVING THE TRACHEA.

As you follow the trachea down the neck to the thorax, you will see that it splits into a Y. Cut the arms of the Y and by pulling, you can peel the trachea away from the esophagus. This is a good time to cut the trachea open and examine the lumen (inside of the trachea).

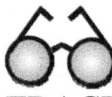

TRACHEAL LUMEN: The lumen should be off-white and smooth with a regular series of ridges and contain nothing but air.

Abnormalities: Red color, plaques, rough areas, foam or other exudate (blood or mucus) in the lumen.

REMOVING THE GASTROINTESTINAL TRACT (GI tract).

First, undermine and isolate the esophagus from the neck. Sever the esophagus near the head and using it as a handle, pull up and begin loosening the attachments of the GI tract to the body. You will see that the GI tract is loosely attached to the body cavity and can be fairly easily removed. When you get to the cloaca (outside opening to the large intestine), try to cut around the cloaca so that the end result is the entire GI tract with the cloaca attached.

HEAD

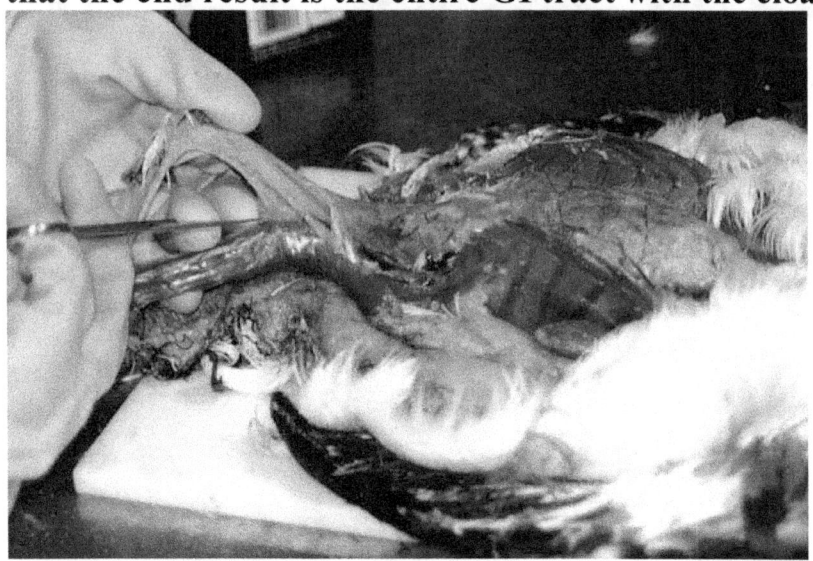

THIS PAGE INTENTIONALLY LEFT BLANK

In this photo, the entire intestinal tract has been extracted and its components labeled. The GI tract is composed of the esophagus, proventriculus, ventriculus, small and large intestines. Cut into portions of each sections to examine the external wall, mucosal surfaces (inside lining), presence of parasites and contents.

ESOPHAGUS: The mucosa should be smooth or slightly wrinkled and light brown to tan.
Abnormalities: Rough sandpaper appearance, plaques, nodules, ulcers, red color or blood.

PROVENTRICULUS: The mucosa should be smooth and light brown to tan. This organ may have a bit of white-tan mucus.
Abnormalities: Rough sandpaper appearance, plaques, nodules, ulcers, red color or blood, worms. NOTE THE CONTENTS

VENTRICULUS: In albatross, this organ is rather small. In ducks and geese, this organ is the gizzard. The main thing to note here are what the contents are.
Abnormalities: Rough sandpaper appearance, plaques, nodules, ulcers, red color or blood, worms. NOTE THE CONTENTS

SMALL AND LARGE INTESTINE: The mucosa and external surface should be smooth and light brown to tan. This organ may contain a bit of white-tan mucus.
Abnormalities: Rough sandpaper appearance, plaques, nodules, ulcers, red color or blood, worms. NOTE THE CONTENTS

PANCREAS: This organ sits in the first loop of the small intestine as it exits the ventriculus. This organ is tan to white and amorphous.
Abnormalities: Excessive enlargement or nodules.

When collecting for formalin fixation, cut segments about 1/2 inch long and DO NOT tie off the ends. When noting contents, indicate if you see any parasites (worms).

PANCREAS

SMALL
INTESTINE

VENTRICULUS

SPLEEN

LARGE
INTESTINE

PRO-
VENTRICULUS

CLOACA

ESOPHAGUS

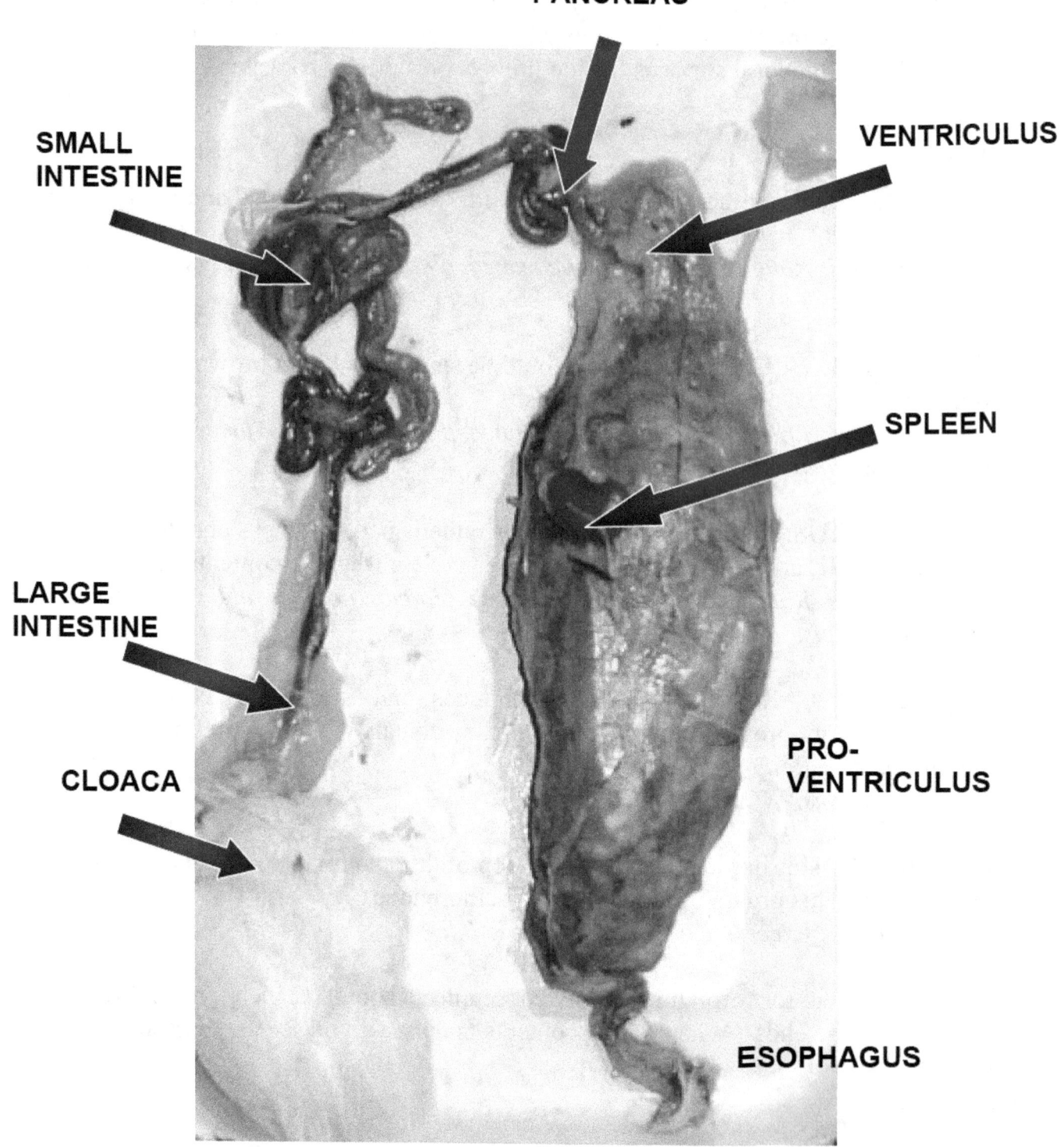

Once you have taken out the GI tract, you are left with the lungs, gonads and kidneys.

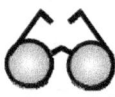

LUNGS: Bird lungs sit tight against the thorax. They should be a uniform light to dark pink.
Abnormalities: Red mottling or uniformly red or grey. Yellow or tan nodules in the lung. When placed in formalin, the lungs should float.

KIDNEYS: The kidneys should be a homogenous red-brown to tan and lobulated. In some cases, the kidneys will be filled with excretion products and will have a pale reticulation throughout.
Abnormalities: Spots, pale color, masses or shrunken size.

GONADS: The gonads sit at the head end of the kidney. The gonads of a mature bird will allow you to differentiate males from females as depicted below. This may not be possible to do in immature birds as the gonads are not developed enough. This is the case in the adjacent photo. In adult birds, the male gonads are roughly bean shaped while the female gonad looks like a cluster of grapes (see below).

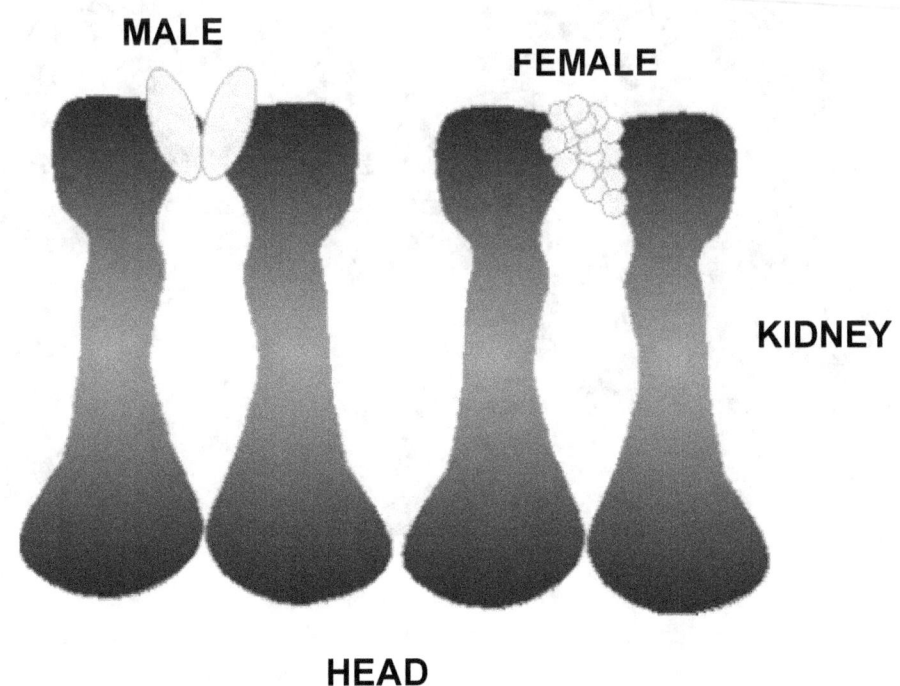

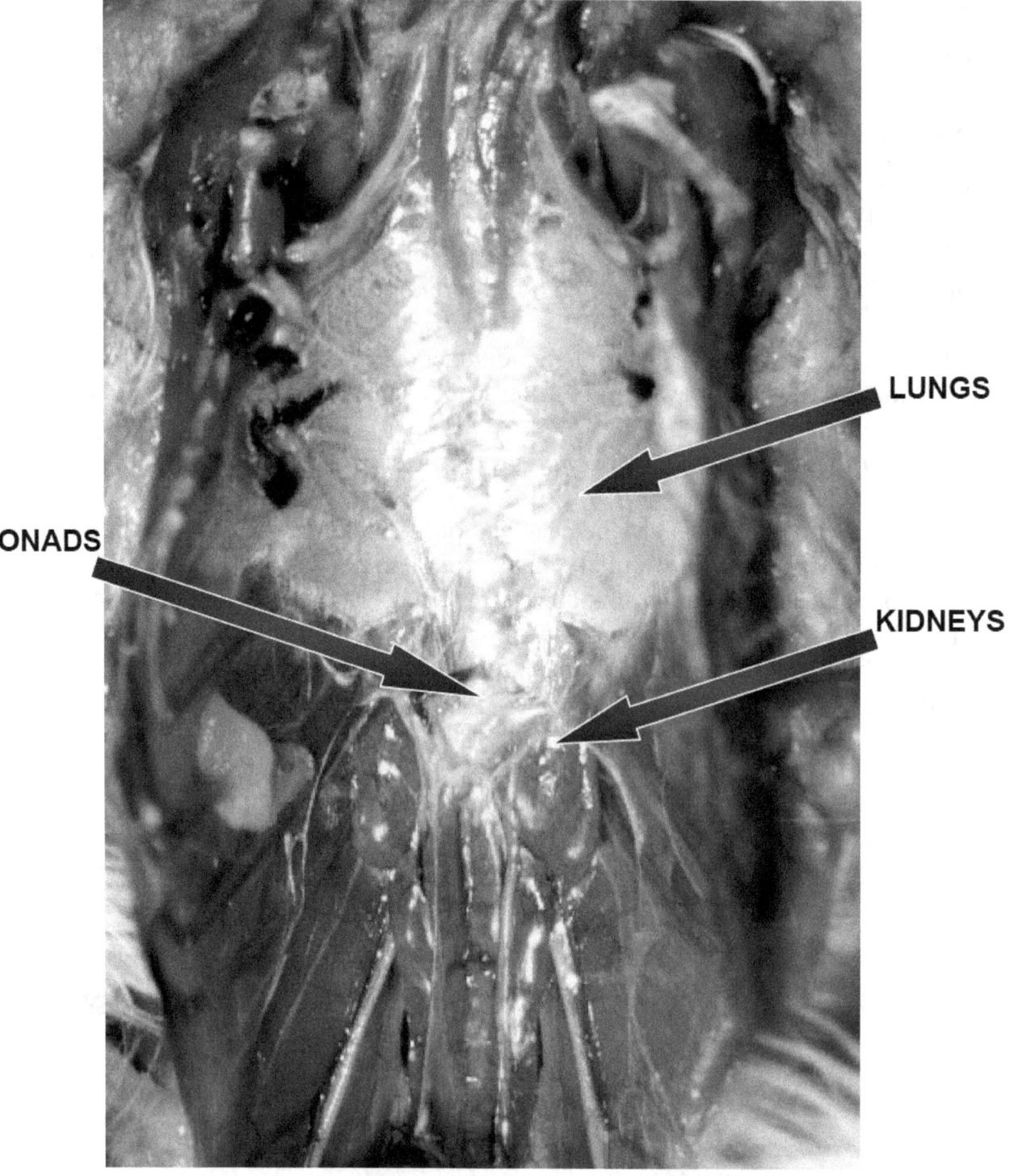

LUNGS

GONADS

KIDNEYS

To remove the lungs, peel them away from the thorax as pictured. This can be most easily done by grabbing the edge of the lung and

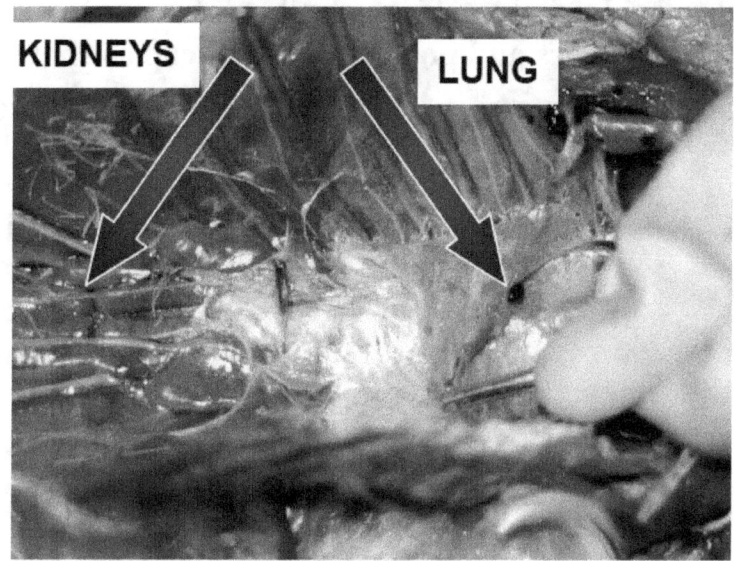

undermining the lung from the thoracic wall as pictured. When you place the lungs in formalin, they should float. NOTE IT if they sink.

To remove the kidneys, grab the cranial pole by the gonads and begin dissection away with scissors underneath. As you gently pull on the kidney, cut away attachments until you have completely removed the organ.

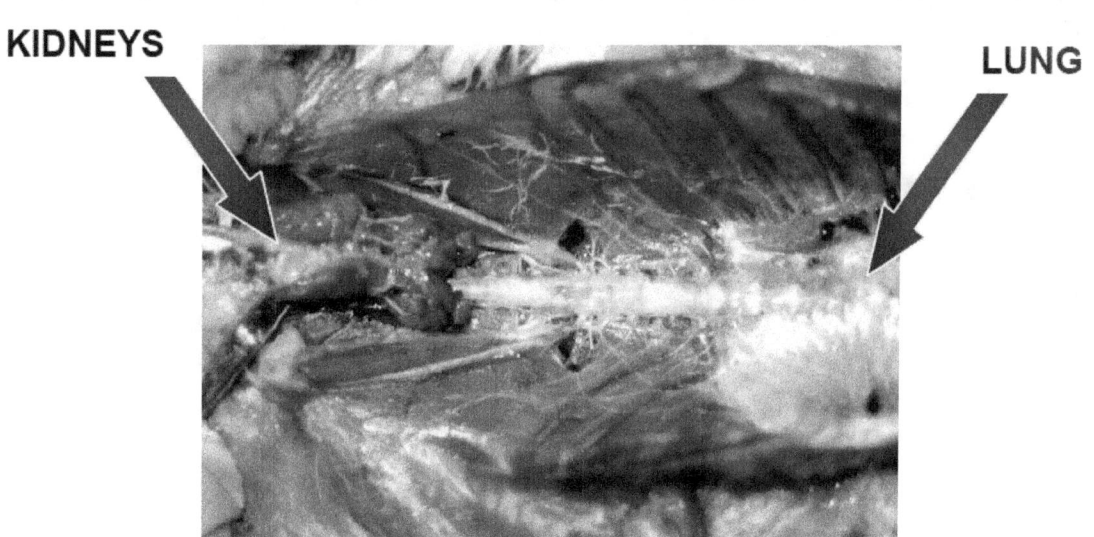

You are almost done. The last step of this process is to remove the brain. To do this you will need to peel the skin off the skull. Cut a crown around the skull with the tips of the shears and pop off the skull cap as shown.

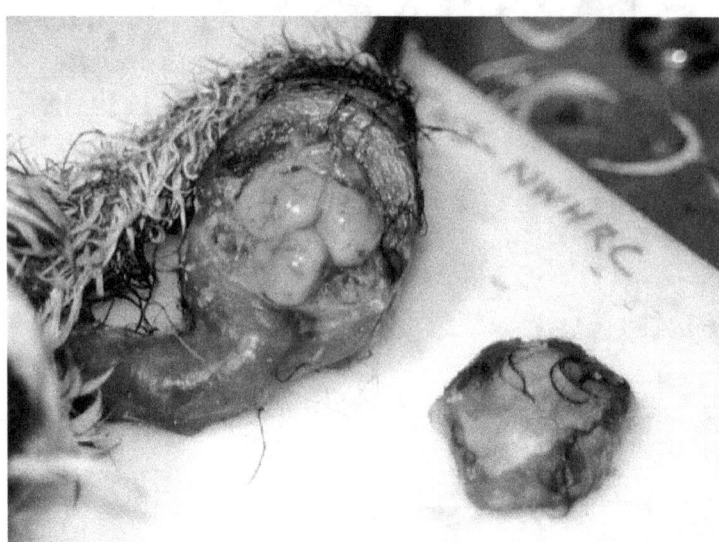

After removing the skull cap, hold the head by the beak, tilt the head upside down, insert your scissors under the front of the brain and cut attachments to the brain. The brain should begin falling out as you cut the nerves attaching it to the skull. Continue dissecting out the brain until it falls to the cutting board.

Ideally, the brain when correctly removed should look something like the picture below. Cut the brain midline down the long axis and place half of it in formalin

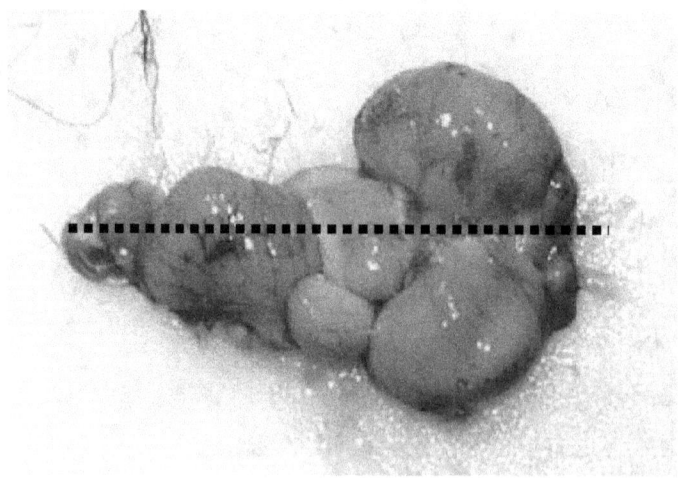

WHEN YOU ARE DONE WITH A NECROPSY, ENSURE THAT:

1) All samples and jars are labeled with a unique number referring to the animal along with date of collection. All organs collected (see p. 28).

2) All information on the necropsy record sheet is complete (see p. 29-30).

3) All "dirty" gloves and other material are disposed of properly. Any sharp items such as scalpel blades and needles should be disposed of in a rigid, sealable puncture proof container (i.e. a plastic jug).

4) Any used formalin is to be stored in sealed containers labeled with the following words: "WARNING: FORMALDEHYDE: HANDLE WITH GLOVES" and disposed of appropriately.

TWO RECIPES FOR 10% FORMALIN

RECIPE 1

If you have graduated cylinders and scale mix the following:

Na_2HPO_4 (Sodium phosphate dibasic)	6.5 g
$NaH_2PO_4.H_2O$ (Sodium phosphate monobasic)	4.0 g
Fresh water	900 ml
37% formaldehyde	100 ml

RECIPE 2

If you have no scales or measuring apparatus

37% formaldehyde	150 ml or 15 parts
Seawater	850 ml or 85 parts

WHEN PREPARING FORMALIN, USE GLOVES AND WORK IN A WELL VENTILATED AREA

CHECKLIST OF ORGANS YOU SHOULD HAVE TAKEN IN FORMALIN

Plain numbers indicate where organ is mentioned in text.
Bold numbers indicate figures in which organs are labeled.

ORGAN	PAGE(S)
TRACHEA	11, **12**
ESOPHAGUS	**11**, 12, **13**, 14, 22
MUSCLE	**12**
LIVER	13, **14, 15, 16**
HEART	13, **14, 15, 16**
THYROID	13, **14, 15**
PROVENTRICULUS	13, **14, 15, 22**
SPLEEN	18, **22**
VENTRICULUS	21, **22**
SMALL INTESTINE	21, **22**
PANCREAS	21, **22**
LARGE INTESTINES/CLOACA	21, **22**
LUNGS	23, **24, 25**
KIDNEYS	23, **24, 25**
GONADS	23, **24**
BRAIN	**26**

NECROPSY DATA SHEET
(all measurements are metric)

Species_____ID#_____ Date Collected_____Necropsied_____

Mmddyy   Mmddyy

Collection site_____ Weight (kg/g)_____

History _____SEX (M/F/U) AGE _____

(Circle most appropriate term(s)). Add notes as you see fit.
BODY CONDITION: (Good, fair, poor)

POST-MORTEM CONDITION: (Fresh dead, ~1 day old, >2 days old)

EXTERNAL EXAM (Skin, mouth, eyes, nostrils, cloaca)

MUSCULOSKELETAL: (*Pectoral muscle atrophy*-None, moderate, severe; *Fat*: firm, soft, jelly-like; *body cavity*-Lots of fluid, small amounts of fluid, no fluid)

LIVER: (*Surface*: smooth, rough, granular, wrinkled; *Consistency*: firm, friable; *Color*: homogenous/mottled, red, black, brown, purple, tan, yellow.)

HEART: (*Surface*: smooth, rough, granular, wrinkled; *Consistency*: firm, friable; *Color*: homogenous/mottled, red, pink, black, brown, purple, tan, yellow.)

LUNGS: (*Surface*: smooth, rough, granular, wrinkled; *Consistency*: firm, friable, spongy; *Color*: homogenous/mottled, pink, tan, yellow, grey, red, brown; **TRACHEA**-*Lumen*: smooth, rough; *Color*: homogenous/Mottled, tan, white, red, brown, green, pink.)

SPLEEN: (*Surface*: smooth, rough, granular, wrinkled; *Consistency*: firm, soft; *Color*: homogenous/mottled, pink, brown, tan, red, black, yellow.)

KIDNEY: (*Surface*: smooth, rough; *Consistency*: firm, soft; *Color*: homogenous/mottled, brown, tan, red, black, brown, yellow.)

GONAD: (*Surface*: smooth, rough; *Consistency*: firm, friable; *Color*: homogenous/mottled, red, black, brown, purple, tan, yellow.)

THYROID: (*Surface*: smooth, rough; *Consistency*: firm, friable; *Color*: Translucent/mottled, orange, red, tan, yellow.)

ORAL: (Mucosa: smooth, rough, granular, pitted; *Color*: homogenous/mottled, pink, tan, yellow, grey, red, brown) Any contents?

ESOPHAGUS-*Mucosa*: smooth, rough; *Serosa*: smooth, rough; *Color*: homogenous/Mottled, tan, white, red, pink.) Contents?

PROVENTRICULUS (*Mucosa*: smooth, rough; *Serosa*: smooth, rough; *Color*: homogenous/mottled, tan, brown, red, yellow, black) Contents?

VENTRICULUS: (*Mucosa*: smooth, rough; *Serosa*: smooth, rough; *Color*: homogenous/mottled, tan, brown, red, yellow, black) Contents?

SMALL INTESTINES: (*Mucosa*: smooth, rough; *Color*: homogenous/mottled, tan, brown, red, yellow, black) Contents?

LARGE INTESTINES: (*Mucosa*: smooth, rough; *Serosa*: smooth, rough; *Color*: homogenous/mottled, tan, brown, red, yellow, black, brown) Contents:

PANCREAS: (*Surface*: smooth, rough; *Consist*: Firm, friable; *Color*: homogenous/mottled, pink, tan, red, yellow, black, brown)

CAECUM: (*Mucosa*: smooth, rough; *Serosa*: smooth, rough; *Color*: homogenous/mottled, tan, brown, red, yellow, black) Contents

SAMPLES:

Formalin:_____

Frozen:_____